Final Weapon

by

Everett B. Cole

Final Weapon
by Everett B. Cole

ISBN: 978-93-59329-64-2

Published by

DOUBLE 9 BOOKS

2/13-B, Ansari Road
Daryaganj, New Delhi – 110002
info@double9books.com
www.double9books.com
Tel. 011-40042856

ABOUT THE AUTHOR

Everett B. Cole (1910-2001) was a professional soldier and a writer of science fiction short tales. He fought at Omaha Beach during WWII and retired in 1960 as a signal maintenance and property officer at Fort Douglas, Utah. He earned a bachelor's degree in math and physics and went on to teach math, physics, and chemistry at Yorktown High School in Texas. In 1951, his debut science fiction story, "Philosophical Corps," appeared in the journal Astounding. Gnome Press published The Philosophical Corps, a reworked version of the story and two others, in 1962. The Best Made Plans, a second novel, was serialized in Astounding in 1959 but never released as a book. He also co-wrote two volumes about the history of south Texas.

District Leader Howard Morely leaned back in his seat, to glance down at the bay. Idly, he allowed his gaze to wander over the expanse of water between the two blunt points of land, then he looked back at the skeletonlike spire which jutted upward from the green hills he had just passed over. He could remember when that ruin had been a support for one of the world's great bridges.

Now, a crumbling symbol of the past, it stubbornly resisted the attacks of the weather, as it had once resisted the far more powerful blasts of explosives. Obstinately, it pointed its rusty length skyward, to remind the observer of bygone conflict—and more.

Together with the tangled cables, dimly seen in the shoal water, the line of wreckage in the channel, and the weed-covered strip of torn concrete which led through the hills, it testified to the arrival of the air age. Bridges, highways, and harbors alike had passed their day of usefulness.

Not far from the ruined bridge support, Morely could see the huge, well maintained intake of one of the chemical extraction plants. He shook his head at the contrast.

"That eyesore should be pulled down," he muttered. "Should have been pulled down long ago. Suggested it in a report, but I suppose it never got to the Old Man. He depends on his staff too much. If I had the region, I'd—"

He shook his head. He was not the regional director—yet. Some day, the old director would retire. Then, Central Coördination would be examining the records of various district leaders, looking for a successor. Then—

He shrugged and turned his attention to his piloting of the borrowed helicopter. It was a clumsy machine, and he had to get in to Regional Headquarters in time for the morning conference. There would be no sense it getting involved in employee traffic—not if he could avoid it.

The conference, his informant had told him, would be a little out of the ordinary. It seemed that the Old Man had become somewhat irritated by the excess privileges allowed in a few of the eastern districts. And he was going to jack everyone up about it. After that would come the usual period of reports, and possibly a few special instructions. Some of the leaders would have pet projects to put forward, he knew. They always did. Morely smiled to himself. He'd have something to come up with, too.

And this conference might put a crimp in Harwood's style. Morely had carefully worded his progress report to make contrast with the type of report that he knew would come from District One. George Harwood had been allowing quite a few extra privileges to his people, stating that it was good for morale. And, during the past couple of months, he'd seemed to be proving his point. Certainly, the production of the employees from the peninsula had been climbing. Harwood, Morely decided would be the most logical person—after himself—for the region when the Old Man retired. In fact, for a time, it had looked as though the director of District One was going to be a dangerous rival.

But this conference would change things. Morely smiled slowly as he thought of possible ways of shading the odds.

He looked ahead. Commuters were streaming in from the peninsula now, to make for the factory parking lots. His face tightened a little. Why, he wondered, had the Old Man decided to call the conference at this hour? He could have delayed a little, until commuter traffic was less heavy. He'd been a district leader once. And before that, under the old government, a field leader. He should know how annoying the employee classes could be. And to force his leaders to mingle with commuting employees in heavy traffic!

For that matter, everyone seemed to be conspiring to make things uncomfortable today. Those heavy-handed mechanics in the district motor pool, for example. They'd failed him today. His own sleek machine, with its distinctive markings was still being repaired. And he'd been forced to use this unmarked security patrol heli. The machine wasn't really too bad, of course. It had a superb motor, and it carried identification lights and siren, which could be used if necessary. But it resembled some lower-class citizen's family carryall. And, despite its modifications, it still handled like one. Morely grimaced and eased the wheel left a little. The helicopter swung in a slow arc.

Helis were rising from the factory lots, to interlace with incoming ships before joining with the great stream headed south. The night workers were heading for home. Morely hovered his machine for a moment, to watch the ships jockey for position, sometimes barely avoiding collisions in the stream of traffic. He watched one ship, which edged forward, stopped barely in time to avoid being hit, edged forward again, and finally managed to block traffic for a time while its inept driver fooled with the controls and finally got on course.

"Quarrelsome, brawling fools," he muttered. "Even among themselves, they can't get along."

He looked around, noting that the air over the Administrative Group was comparatively free of traffic. To be sure, he would have to cross the traffic lines, but he could take the upper lanes, avoiding all but official traffic. A guard might challenge, but he could use his identifying lights. He wouldn't be halted. He corrected his course a little, glanced at the altimeter, and put his ship into a climb.

At length, he eased his ship over the parklike area over Administrative Square and hovered over the parking entry. A light blinked on his dash, to tell him that all the official spaces were occupied. He grunted.

"Wonder they couldn't leave a clear space in Official. They know I'm coming in for conference."

He moved the control wheel, allowing his ship to slide over to a shopping center parking slot, and hovered over the entry, debating. He could park here and take the sub-surface to Administrative, or he could use the surface lot just outside of the headquarters group. Of course, the director frowned on use of the surface lot, except in emergency. The underground lots were designated for all normal parking. Morely thought over the problem, ignoring the helis which hovered, waiting for him to clear the center of the landing area. Finally, his hand started for the throttle. He would settle in the landing slot, let the guards shove his heli to a space, and avoid any conflict with the director's orders regarding the surface lot.

Suddenly, there was a sputtering roar. Someone had become impatient at the delay. A small sports heli swept by, impellers reversed, and dropped rapidly toward the entry to the underground parking space. Morely's ship rocked a little in the air blast.

For an instant, Morely felt a sharp pain which gnawed at the pit of his stomach. His head was abruptly light, and his hand, apparently of its own volition, closed over the throttle knob.

This joy boy was overdue for a lesson.

Morely measured the distance quickly, judging the instant when the other pilot would have to repitch his impellers and halt his downward rush. He allowed his own heavy ship to wallow earthward.

Scant feet from ground surface, the sportster pilot flicked his pitch control and pulled his throttle out for the brief burst of power which would allow him to drop gently to the landing platform.

Morely grinned savagely as he saw the impellers below him change pitch and start to move faster. He twisted his own impellers to full pitch and pulled out the throttle for a sudden, roaring surge of power, then swung the control column, jerking his ship up and away. As he steadied his heli and cut power, he looked down.

The powerful downblast had completely upset the sportster pilot's calculations. The small ship, struck by the gale from above, had listed to the right and gone out of control, grazing one of the heavy splinter shutters at the side of the landing slot. The ship lay on its side, amidst the wreckage of its impellers.

Morely flicked on his warning siren and lights, then feathered his own impellers, dropping his ship in free fall. He dropped to the grassy area by the landing slot, ignoring the other ships which scattered like frightened chickens, to give him room. At the last instant, he twisted the impellers to full pitch again, pulled out the throttle for a moment, then slammed the lever to the closed position. His ship touched down on springy turf, its landing gear settling gently to accept the weight. A klaxon was sounding, and warning lights flashed from the landing slot, to warn ships away from an attempted landing.

It would be a long time before the shiny, new sportster would be in condition to sweep into another parking area. And, after paying his fine and taking care of his extra duties, it would be an even longer time before the employee-pilot would have much business in the luxury shopping center, anyway.

Morely smiled bitterly as he closed the door of his ship. It didn't pay to cross Howard Morely—ever.

He walked slowly toward the landing slot, motioning imperiously to an approaching guard.

"Have someone place that ship for me," he ordered, jerking a thumb back toward his heli. "Then come over to that wreck. I shall want words with the pilot." He held out his small identification folder.

The guard's glance went to the folder. For an instant, he studied the card exposed before him, then he straightened and saluted, his face expressionless.

"Yes, sir." He signaled another guard, then pointed toward Morely's ship, and to the landing slot. "I can go with you now."

The two went down in the elevator and walked over to the wrecked sportster. A slender man was crawling from a door. When the man was clear of his ship, Morely beckoned.

"Over here, Fellow," he commanded.

The sportster pilot approached, the indignation on his face changing to bewilderment, then dismay as he noted Morely's insignia and the attitude of the two men who faced him.

Morely turned to the guard.

"Get me his name, identification number, and the name of his leader."

"Yes, sir."

The guard turned to the man, who grimaced a little with pain as he slowly put a hand in his pocket. Wordlessly, he extracted a bulky folder, from which he took a small booklet. He held out the booklet to the guard.

Morely held out a hand. "Never mind," he said. "Simply put him in custody. I'll turn this over to his leader myself."

He had noted the cover design on the booklet. It was from District One—Harwood's district. He flipped the cover open, ascertaining that there was no transfer notice. He'd give this to Harwood all right—at the right time. He looked at his watch.

"I shall want my heli in about three hours," he announced. "See to it that it's ready. And have a man check the fuel and see if the ship's damaged in any way." He turned away.

The district leaders sat before the large conference table. Among them, close to the director's place, was Morely, his face fixed in an expression of alert interest. His informant had been right. The man must have gotten a look at the Old Man's notes. The regional director was criticizing the laxity in inspection and control of employee activities. He objected to the excessive luxury activity allowed to some members of the employee classes, as well as to the overabundance of leisure allowed in several cases, some of which he described in detail.

He especially pointed up the fact that a recent heli meet had been almost dominated by employee class entries. And he pointed out the fact that there was considerable rehabilitation work to be done in bombed areas. It could be done by employees, during their time away from their subsistence jobs. That was all community time, he reminded.

It was all very well, he said, to allow the second- and even third-class citizens a certain amount of leisure recreation. That kept morale up. But they were certainly not to be allowed any position of dominance, either individually, or as a class. That, he said, was something else again. It was precisely the sort of thing that had led to the collapse and downfall of many previous civilizations.

"Keep 'em busy," he ordered. "So busy they don't have time to think up mischief to get into. Remember, gentlemen, second- and third-class citizens have no rights—only privileges. And privileges may be withdrawn at any time."

He rapped sharply on the table and sat down, looking at the leader of District One.

One by one, the district leaders made their verbal reports of activity. Occasionally, questions of production or work quotas were brought up and decided. Morely waited.

At last, he made his own report, emphasizing the fact that his district had exceeded its quotas—subsistence, luxury, and rehabilitation—for the fourth consecutive quarter. He cited a couple of community construction projects he had ordered and which were well on the way to completion, and brought out the fact that his people, at least, were being inspected constantly and thoroughly.

Also, he suggested, if any time remained to be used, or if leisure activity threatened to become excessive, it might be well to turn some attention outside of the old urban areas. There was considerable bomb damage in the suburban and former farming areas, and the scrap from some of the ruined structures could be stockpiled for disposal to factories and community reclamation plants.

Further, a beautification program for the entire region might keep some of the employee class busy for some time. And some of the ex-farmers among the lower classes might find it pleasant to work once again with the soil, instead of their normal work in the synthetic food labs or machine shops. With the director's permission, he could start the program by removing the useless tower and wreckage at the bay channel, and by salvaging the metal from it. Of course, he admitted, it was a trifle beyond his own authority, since most of the channel was in District One. The regional director cast him a sharp glance, then considered the suggestion. At last, he nodded.

"It might be well," he decided. "Go ahead, Morely. Take care of that detail." He looked over at his executive. "Have Planning draw up something on salvage and beautification in the former rural areas," he ordered. He looked about the room.

"And the rest of you might try looking over your own districts. You don't have to wait for a directive, and every one of you can find some improvement that could be made. If it's a district line matter, submit some plan for mutual agreement to my office." He rose and went to the door.

Morely waited, watching George Harwood. The leader of District One gathered his papers, looked down the table for an instant, then went out. Morely followed him at a discreet distance.

As Harwood neared the door to the regional director's office, Morely caught up with him.

"Oh, Harwood," he said loudly. "Caught one of your people in a flagrant case of reckless flying this morning. Why don't you bear down a little on those fellows of yours? This one seemed to think he was winning a heli meet."

He held out the folder he had confiscated. "Here's his identification. I had the guards hold him for you. Second-class citizen. Must've had a lot of spare time, to get the luxury credits and purchase authorization for that ship of his."

Harwood looked at him, a faint expression of annoyance crossing his face. Then, he glanced at the open door nearby, and comprehension grew on his face. He took the folder, nodded wordlessly, and walked rapidly past Morely, who turned to watch him.

As Harwood swung through the door to an elevator, Morely smiled appreciatively. That had been a smart trick, he thought. Have to remember that one. No argument to disturb the Old Man. Not even positive proof that Morely hadn't been talking to empty space. But there was an answer to that, too, if one was alert. He walked through the doorway into the director's office.

The regional director looked up.

"Oh, Morely. You wanted to see me?"

"Yes, sir." Morely stood at rigid attention. "I just thought of all those useless highways around the countryside. Of course, a few of them have been camouflaged and converted to temporary and emergency heli parking lots, but there's still a lot of waste concrete about that could be removed. It would improve the camouflage of the groups. It could be divided into community projects for spare time work, sir."

"Very good idea. If this stalemate we're in should develop into another war, it would be well to have as few landmarks as possible. And some of these people do have too much time on their hands. They sit around, thinking of their so-called rights. Next thing we know, some of the second-class citizens'll be screaming for the privilege of a vote. Set it up in your district, Morely. We'll see how it works out, and the rest of the district leaders can follow your example."

He looked sharply at Morely. "Heard a little disturbance in the hall just before you came in."

"Oh, that." Morely contrived a look of confusion. "I'm sorry, sir. I didn't mean anyone to hear that. It was just that I had a minor bit of business with Leader Harwood. One of his people nearly knocked me out of the air this morning, over a parking area, and I confiscated his identification. I tried to give it to Harwood after the conference, but he must have been in a hurry. I caught up with him and gave him the folder."

"So I heard." The director smiled wryly. "Anything more?"

"No, sir." Morely saluted and left.

"That," he told himself, "should drop Harwood a few points."

He went to the parking area to reclaim his helicopter. Better get back to his district and start setting up those community projects. Too, he would have to run a check inspection or so this evening. See to it his sector men weren't getting lax. He'd check on Bond tonight.

He flew back to District Twelve, dropped his helicopter into the landing area, and made his way to his office.

Inside, he went to a file, from which he took his spot-inspection folder. Carrying it to his desk, he checked it. Yes, Bond's sector was due for a spot inspection. Might be well to make a detailed check of one of the employees in that sector, too. Morely touched a button on his desk.

Almost immediately, a clerk stood in the doorway.

"Get me the master quarters file for Sector Fourteen," Morely ordered.

The clerk went out, to return with two long file drawers. Quickly, he set them side by side on a small table, which he pushed over to his superior's desk.

Idly, Morely fingered through the cards, noting the indexing and condition of the file. He nodded in approval, then gave the clerk a nod of dismissal. At least, his people were keeping their files in order.

He reached into a pocket, to withdraw a notebook. Turning its pages, he found a few of the entries he had made on population changes, then cross-checked them against the files. All were posted and properly cross-indexed. Again, he nodded in satisfaction.

Evidently, that last dressing down he had given the files section had done some good. For a moment, he considered calling in the chief clerk and complimenting him. Then, he changed his mind.

"No use giving him a swelled head," he told himself.

He drew a file drawer to him, running his finger down its length. At last, he pulled a card at random. It was colored light blue.

He put it back. Didn't want to check a group leader. He'd be a first-class citizen, and entitled to privacy. He pulled another card from a different section of the file. This one was salmon pink—an assistant group leader. He examined it. The man was a junior equipment designer in one of the communications plants. For a moment, Morely tapped the card against his desk. Actually, he had wanted a basic employee, but it might be well to check one of the leadmen. He could have the man accompany him while he made a further check on one of the apartments in his sub-group. Again, he looked at the card.

Paul Graham, he noted, was forty-two years of age. He had three children—was an electronics designer, junior grade. His professional profile showed considerable ability and training, but the security profile showed a couple of threes. Nothing really serious, but he would be naturally expected to be a second-class citizen—or below. It was not an unusual card.

Morely looked at the quarters code. Graham lived in Apartment 7A, Group 723, which was in Block 1022, Sector Fourteen. It would be well to check his quarters first, then check, say, 7E. Morely went through the numerical file, found the card under 7E, and flipped the pages of his notebook to a blank sheet, upon which he copied the data he needed from the two cards.

He put the notebook in his pocket and returned the cards to their places in the file, then riffled the entire file once more, to be sure there would be no clue as to which cards he had consulted. Finally, he touched the button on his desk again.

Once more, the clerk stood in the doorway.

"This file seems to be satisfactory," he was told. "You may bring in the correspondence now."

The correspondence was no heavier than usual. Morely flipped through the routine matter, occasionally selecting a report or letter and abstracting data. Tomorrow, he could check performance by referring to these. At last, he turned to the separate pile of directives, production and man-hour reports, and other papers which demanded more attention than the routine paper.

He worked through the stack of paper, occasionally calling upon his clerk for file data, sometimes making a communicator call. At last, he pushed away the last remaining report and leaned back. He spun his chair about, activated the large entertainment screen, and spent some time watching a playlet. At the end of the play, he glanced at his watch, then turned back to his desk. He leaned forward to touch a button on his communicator.

As the viewsphere lit, he flicked on the two-way video, then spoke.

"Get me Sector Leader Bond." He snapped the communicator off almost before the operator could acknowledge, then spun about, switching his entertainment screen to ground surface scan. A scene built up, showing a view from his estate in the hills.

There were some buildings on the surface—mostly homes of upper grade citizens, who preferred the open air, and could afford to have a surface estate in addition to their quarters in the groups. These homes, for the most part, were located in wooded areas, where their owners could find suitable fishing and hunting.

Most of the traces of damage done by the bombings of the Nineties were gone from about the estate areas by now, and the few which remained were being eliminated. Morely increased the magnification, to watch a few animals at a waterhole. He could do a little hunting in a few weeks. Take a nice leave. He drew a deep breath.

Those years after the end of the last war had been hectic, what with new organizational directives, the few sporadic revolts, the integration of homecoming fighters, and the final, tight set-up. But it had all been worth it. Everything was running smoothly now.

The second- and third-class citizens had learned to accept their status, and some few of them had even found they liked it. At least, now they had far more security. There was subsistence in plenty for all producers, thanks to the war-born advances in technology, and to the highly organized social framework. To be sure, a few still felt uneasy in the underground quarters, but the necessity for protection from bombing in another war had been made clear, and they'd just have to get used to conditions. And, there were a very few who, unable to get or hold employment, existed somehow in the spartan discomfort of the subsistence quarters.

For most, however, there was minor luxury, and a plenitude of necessities. And there was considerable freedom of action and choice as well as full living comfort for the full citizens, who had proved themselves to be completely trustworthy, and who were deemed fit to hold key positions.

The communicator beeped softly, and he glanced at the sphere. It showed the face of Harold Bond, leader of the fourteenth sector. The district leader snapped on his scanner.

"Report to me here in my office at eighteen hours, Bond."

"Yes, sir."

"And you might be sure your people are all in quarters this evening."

Bond nodded. "They will be, sir."

"That's all." Morely flicked the disconnect switch.

He got up, strode around the office, then consulted his watch. There would be time for a cup of coffee before Bond arrived. Time for a cup of coffee, and time for the employees in Sector Fourteen to scurry about, getting their quarters in shape for an inspection. They would have no way of knowing which quarters were to be checked, and all would be put in order.

He smiled. It was a good way, he thought, to insure that there would be no sloppiness in the homes of his people. And it certainly saved a lot of inspection time and a lot of direct contact.

He went out of the office, and walked slowly down to the snack bar, where he took his time over coffee, looking critically at the neat counter and about the room as he drank.

The counter girls busied themselves cleaning up imaginary spots on the plastic counter and on their equipment, casting occasional, apprehensive glances at him. Finally, he set his cup down, looked at the clock over the counter, and walked out.

Bond was waiting in the office. Morely examined the younger man, carefully appraising his appearance. The sector leader, he saw, was properly attired. The neat uniform looked as if freshly taken from the tailor shop. The man stepped forward alertly, to halt at the correct distance before his superior.

"Good evening, sir. My heli is on the roof."

"Very good." Morely nodded shortly and took his notebook from his pocket. "We'll go to Building Seven Twenty-three."

He turned and walked toward the self-service elevator. Bond hurried a little to open the door for him.

Bond eased the helicopter neatly through the entry slot and on down into one of the empty visitor spaces in the landing area at Block 1022. The two men walked across the areaway to an entrance.

As they went up the short flight of stairs into the hall, Morely took careful notice of the building. The mosaic tile of the stairs and floor gleamed from a recent scrubbing. The plastic and metal handrails were spotless. He looked briefly at his subordinate, then motioned toward the door at their right.

"This one," he ordered.

Bond touched the call button and they waited.

From inside the apartment, there was a slight rustle of motion, then the door opened and a man stood before them. For an instant, he looked startled, then he straightened.

"Paul Graham, sir," he announced. "Apartment 7A is ready for inspection." He stepped back.

Morely looked him over critically, saw nothing that warranted criticism, and went inside, followed by Bond.

Cursorily, the district leader let his gaze wander about the apartment. The kitchen at his left, he saw, was in perfect order, everything being in place and obviously clean. He went to the range and motioned with his head.

"Pull the drip pan," he ordered.

Graham came forward and pulled a flat sheet from the range, then opened an access door at the front of the stove.

Morely peered inside, then thrust a hand in. For a moment, he groped around, then he pulled his hand out and looked at it. It was clean. He sniffed at his fingers, then turned away.

"You may replace the pan, Fellow." He went into the living room, noting that the woman and three children were neat and in the proper attitudes of attention. One of the children was looking at him, wide-eyed. He saw that the child was clean and apparently healthy.

In addition to the usual chairs, table, and divan, there were some bookcases which formed a small alcove around a combination desk and drawing table. Morely circled the bookcases, to stand before the desk.

"What's this?" he demanded. He turned to a bookcase, to examine the titles.

Most of the books were engineering texts and reference works. There were some standard works of philosophy and a few on psychology. None of the titles seemed to be actually objectionable.

"I—" Graham started to speak, but Morely silenced him with an upraised hand.

"Later," he said coldly. "Bond, has this been reported to you, and have you investigated?"

Bond nodded. "Yes, sir," he said. "Graham is a design engineer, sir, and has been granted permission to do some research in his quarters.

"He's commercially employed, sir, and it was a routine matter. His employer says he has been keeping his production quotas, no alteration to the apartment has been made, and no community property has been defaced. I'm told that several of Graham's designs have been of value in his plant. I didn't think—"

"I see you didn't. What is this man working on now?"

"A new type of communicator, sir. I don't know all the details."

"Get them, Bond. Get them all, and give me a full report on his project and its progress tomorrow. Since this work is being done during time when the man is not working for his employer, he's using community time and the community becomes vitally interested in his results." Morely paused, looking at the bookcase again.

"And, while we are on the subject," he added, "get me details on those previous designs you spoke of. It's quite possible the community has not been getting royalty payments to which it's entitled." He picked out a book, flipping over its pages for a moment, then replaced it and looked searchingly at Bond.

"And get me a full inventory of this man's books and any equipment he may have." He turned on Graham.

"Do you have purchase authorization and receipts for all of this?"

"Yes, sir." Graham motioned toward the desk.

"Very well. I shan't bother with that now. An investigating team can check that."

Morely took a final glance at the half-finished schematic on the drawing board, then circled the bookcases again, to come out into the main room.

"We'll inspect the rest of your quarters."

At last, Morely left the quarters area, followed by Bond. As they reached the helicopter, Morely turned, one hand on the door.

"Laxity, Bond, is something I don't tolerate. You should know that. Possibly this man, Graham, is doing nothing illegal, or even irregular. Possibly, he is not wasting community time, but I have very serious doubts. I'll venture to say the community has a financial interest in several of his recent designs, and I mean to find out which ones and how much. And it's certainly an unusual situation. The man's a leadman, you know, and could spend his time more profitably in checking on the people he's responsible for." He slid into the seat.

"I'll concede," he continued, "that employees are to be allowed a certain amount of recreation of their own choosing. They may have light reading in their quarters, and they may even work on small projects—with permission, of course. But this man seems to have gone much farther than that. He has a small electronics factory of his own, as well as a rather extensive library. He's obviously spending a lot of time at his activities, and that time must come out of his community performance. This certainly is not routine, and I can't condone your failure to make a report on it."

"But, I—"

Morely held up a hand sternly. "Let's not have a string of excuses," he said. "Give me a full report on the man's possessions, his history, and the progress of whatever work he's doing in that private factory of his. Get the details on his previous designs, too. And bring your report in to me in the morning, personally. I shall want to determine whether to make this new device a community project, or whether to allow it to be offered to his employer on a community royalty agreement. And I shall require details on his older designs for Fiscal to examine into. Research, you should know, is a community function, not something to be done in any set of quarters. I shall want to talk to you further when I've gone over this matter.

"Now, get me back to the district offices. I want to get home, and you've work to do tonight."

The report was a long one. Morely smiled to himself as he thought of the time it must have taken Bond to assemble the data and to make up his final draft. Possibly in the future, that young man would be a little less inclined to assume too much authority, or to be too soft in his dealings with the employee classes. The spring in his swivel chair twanged musically as the district leader leaned back to read.

First, there was an inventory of Graham's effects. It was a lengthy list, followed by a certification by a security inspector that all of the equipment inventoried was covered by authorizations and receipts held by Graham, and that none of the books and equipment were of improper nature for possession by a member of the employee classes. Morely grunted and tossed that section aside.

There was a detailed history of Graham's activities, so far as known to Security. Morely scanned through it hurriedly. There was nothing here of an unusual nature.

Graham had been graduated from one of the large technical colleges during the early nineties. Morely noted that it was one of those schools which had been later closed as a result of one of the post-war investigations.

The subject had been employed by Consolidated Electronics as a junior engineer, and had designed several improvements for Consolidated's products. There was a record of promotions and a few awards. He had held a few patents, which had been taken over by the Central Coördination Products Division during the post-war reorganization. He had also belonged to the now proscribed Society of Electronic Engineers, had contributed articles to that organization's journal, and had taken an active part in some of its chapter meetings.

During the war, he had worked on radio-controlled servos, doing acceptable work. When the professional and trade societies and other organizations were outlawed, he had promptly resigned from his society, and made the required declarations. But he had been reported as privately remarking that it was "a sad thing to see the last vestiges of personal freedom removed."

Morely pursed his lips. Not an unusual history, he decided. Of course, the man was completely ineligible for full citizenship—bad risk. He was barely qualified for second-class citizenship, his obvious ability being the only qualifying factor. Unlike many, he had no record of any effort to shirk duty, or do economic damage during the critical period. The district leader tossed the dossier aside and picked up the report on Graham's present activities.

There were a series of complex schematics, and several machine drawings which he shuffled to the back of his report. Those could be interpreted later, if necessary. He was interested in the description of function.

The device Graham was working on was described as a communicator which operated by direct mind-to-mind transfer. Morely sat up straighter, reading the paragraph over again. Either this man was a true genius, who had discovered a new principle, or he was completely a crackpot.

"Telepathy!"

Morely snorted and went over to the descriptions of the device, reading carefully. Finally, he read the comments of a senior engineer, who cautiously admitted that the circuits involved, though highly unconventional, were not of a type to cause spurious radiation, or to interfere with normal communication in any way.

The engineer also noted that it was possible that the device might be capable of radiation effects outside of the electromagnetic spectrum, and

that the power device was capable of integration into standard equipment—
in fact, might be well worth adoption. He carefully declined, however, to
give any definite opinion without an actual model to run tests on. And he
added the comment that the first model was as yet incomplete.

Morely tossed the last sheet to his desk and leaned forward, tapping
idly on the dull-finished plastic. Finally, he touched his call button and
waited till the clerk came in.

"You may send Mr. Bond in now," he directed.

He picked up the section of the report dealing with Graham's past
designs, and started scanning it. He would have the Fiscal chief go over this
and set up the necessary royalty agreements with Consolidated. Some of
them might generate worth-while amounts of funds.

He made no sign of recognition or awareness when Bond entered the
office, but continued with his reading. At last, he pulled a notepad to him,
wrote a brief indorsement to the Fiscal chief, and clipped it to the part of the
report dealing with Graham's older designs. He replaced his pen in its stand
and leaned back, to stare at his junior, who stood at rigid attention.

"Yes?"

"Sector Leader Bond, sir, reporting as ordered." Bond saluted.

Negligently, Merely returned the salute, then picked up Bond's report.

"I have gone through this, Bond," he announced. "Very interesting.
And you thought it too unimportant to report on before?"

"I didn't want to bother you with some idle fantasy, sir. Until the man's
experiments showed definite results of some sort, I—"

"And then, you hoped to spring a completed device on me? Take credit
for it yourself, eh?"

"Not at all, sir. I—"

Morely raised a hand. "Never mind. I don't need any kind of aid to read
your intentions. They're quite plain, I see. It would have been quite a credit
to you, wouldn't it?

"'Look what I worked out, with a little, minor help from one of the
employees in my sector.'

"But I've seen that line worked before, Bond, and worked smoothly.
You don't catch the Old Man napping so easily as that." He paused.

"Of course we don't know whether or not this device is going to be of
any real use. But we do know that this man, Graham, has developed one

thing which can be profitably incorporated into conventional equipment. That power source of his appears to be quite practical, and we'll adopt it. Offer it to the man's employer, subject to community royalty. And see if you can get Graham a little time off work in compensation. Then, keep a close watch on his work on the rest of his device. He'll probably use his time off to work on it—at least, he'll be a lot better off if he does.

"I want frequent reports on his progress—daily reports, if any significant developments occur. And I want a model of that device as soon as it's developed and has had preliminary tests. If it works, it might be valuable for community defense." He waved a hand.

"That's all."

Bond turned to go, and almost got to the door before Morely called him back.

"Oh, one more thing, Bond. Keep a closer watch on the rest of your people. If any more of them decide to do extra work of any unusual nature, I shall expect an immediate report in full. Don't fail me again. Is that clear?"

"Yes, sir." Bond saluted again and made his escape.

Morely watched him disappear, then turned to his communicator. "Get me Field Leader Denton," he ordered.

The pause was slight, then the face of a middle-aged man appeared in the viewsphere.

"Denton," said the district leader, "I want you to keep closer watch on your sector men. Last night I spot-checked Bond, in Fourteen, and I found an irregularity. I'll expect you to endorse the report back, and I'll expect you to tighten down. Keep an especially close eye on this man, Bond."

The field leader's eyebrows raised a little. "Bond, sir? He's one of—"

"Bond. Yes." His superior interrupted forcefully. "And tighten down on all your men. You know how I feel about laxity."

He snapped the communicator off and gathered Bond's report together. For a few seconds, he looked at the neat stack of paper, then he slipped a paper clamp on it and punched his call button.

"There!" Paul Graham straightened from his hunched-over position at the desk. He laid his soldering iron down and massaged the small of his back, grimacing slightly.

"Oh, me! I'll swear my back'll never be the same again. But that ought to do it, at last." He looked at the equipment before him and grinned ruefully.

"Of all the haywire messes. It started out so nice. And it ended up so awful."

The device *had* started out as a fairly neat assembly, using a headband as a chassis. But the circuitry seemed to have gone out of control. Miniature sub-assemblies hung at all angles from their wires and tiny components were interlaced through the unit, till the entire assembly looked like a wig from a horror play. Graham shook his head, picked up the band; and carefully fitted it, being careful that the contacts touched his forehead and temples properly.

For an instant, he looked a little dazed. Then, he reached up and fumbled for a moment with the controls at the front of the headband. Suddenly, he stopped, an expression of pleasure on his face. He stood for a time, looking at the wall, then looked up at the ceiling. He frowned and looked at his wife, who was anxiously watching him. A smile grew on his face, and she was clearly conscious of the projected thought.

"I told you, Elaine, it can't possibly hurt anyone. Stop worrying about me."

Elaine Graham looked startled. "I didn't, say anything, darling."

Her husband looked at her with an impish grin. She frowned a little, then her eyes widened and her mouth opened a little. She ran at him indignantly.

"It simply isn't decent! You take that thing off, Paul Graham, right now. I won't have you reading my mind!"

Graham laughingly fended her off with one hand as he carefully removed the headband with the other. He set the device gently on the desk, then seized his wife about the waist.

"It works, honey," he said jubilantly. "It really works." He waltzed her away from the desk, to the middle of the living room.

"Of course, I couldn't get anything from anyone but you. It seems to work just as I thought it might—only if you can see the person you want to contact. But I'll bet two people who were acquainted could use two of these things to communicate with each other at any distance. And it may be possible to work out the problem of single-device communication at distance and through obstacles. But that'll have to come later. Right now, this thing works."

"But Paul. I'm afraid. What will *they* do with something like this? We have so little freedom left now. Why, they won't even let us think privately." She paused, her head turning from side to side as she looked about the apartment.

"You know, Paul, I hardly ever dare go out of this apartment now, they upset me so. And if they're able to read my thoughts, I shan't be safe, even here."

Graham frowned. "True," he admitted. "But somehow, when I had the thing on, I got some funny ideas. I wonder if anyone could really oppress someone he fully understood. I wonder if two people who could fully comprehend each other's point of view could have a really serious disagreement." He picked up the headband, looking at it searchingly.

"And there's another thing," he added. "Unless both parties are wearing the things, vision seems to be essential to any reaction, at least in this model. I tried to get thoughts from the kids and from the Moreno's, upstairs. But there wasn't a thing. And yet, I could get you clearly. Apparently this thing won't work out as a spy device."

"But, are you sure?"

Graham shrugged wryly. "Well ... no," he admitted. "I'll have to finish wiring the other set and try 'em both out before I'll be sure of anything. And it'll take a lot of tests before I'm sure of very much. Now, I've just got some ideas." He frowned thoughtfully.

"Anyway, I can't stop now. They know about the thing, and I've got to finish it—or furnish definite proof it's impractical." He turned back to the desk. "Should be through with the other band in a few minutes. Just have to put in a couple of filters."

He picked up the completed device and turned around again. "Here, Elaine, put this on, will you? See what you get. Try to catch a thought from outside the room."

Dutifully, Elaine Graham accepted the headband. She eyed it doubtfully for a moment, then adjusted it over her hair, setting the contacts on her skin as she had seen her husband do. For a few seconds, she stared at her husband, wide-eyed. Then, she looked away, her eyes focused on infinity.

Graham busied himself with the soldering iron and another headband.

At last, Elaine took the headband off. "It's weird, Paul," she said. "When I was looking at you, I knew everything you were thinking. But when I looked away, there was nothing. It was almost as though I didn't have it on. Only, I seemed to be able to think so much more clearly."

Graham looked up from his work, squinting thoughtfully. "Yeah," he muttered. "Yeah, I noticed that, too, come to think of it. Feedback effect of some sort, I suppose. Have to experiment with that, too, I expect." He turned back to his work.

Elaine put the headband back on and watched him. She felt a complete familiarity with everything he was working on. For the first time, she felt she fully understood this man with whom she had lived for so many years. And the understanding was pleasant. She could comprehend the mysteries of the circuits he was working on. She had always felt slightly neglected when he worked with his equipment, especially since the bureaucracy, who took his results without recompense. Now, she could feel his interest in his work for its own sake. She could sympathize with it. And, with a little study, she felt she could join with him.

Graham straightened again. "It's done," he said. He picked the second headband from the desk and put it on. Abruptly, both he and his wife were aware of a fuzziness in their thoughts and senses. The walls, the floor, and the furniture seemed to blur and waver, like the fantasy world of delirium. He put his hand up and adjusted the controls. The room returned to normal, and their senses were abruptly sharp and clear again. He dropped his hand.

"Outside. See if it'll work when we can't see each other."

"Almost curfew time."

"Only a couple of minutes. Then lights out and sleep."

Elaine walked to the door. She stepped out into the corridor and walked down the steps.

"All right?"

"Perfect! Try the parking lot. Close the door."

She went out of the quarters, crossed the areaway, and stood under the landing slot. Far overhead, a segment of sky appeared between the open bomb shutters. Stars shone coldly. She was conscious of a movement and looked down, toward a shadow which moved among the parked helicopters.

"What's that?"

She looked more closely at the shadow, then shuddered a little.

"Never mind." The thought was urgent. *"Come inside. I got him, too."*

Quickly, Elaine walked back into the apartment. She closed the door and walked to the desk, removing the headband as she approached. Her husband put his headband beside it.

"We'd better get to bed," he said quietly. "I'll notify them tomorrow."

"No, Paul. It would be harder then. And there would be so many questions. Call the sector leader tonight. We'll have to get it over." Elaine shivered.

"But what *will* they do with it?" She asked the question almost despairingly.

Graham shook his head. "I'm not sure," he admitted. "I started with the idea of simply building a really effective communicator. But this is more than that. To you and I, it meant full understanding. But to that person out there ... I don't know."

"His thoughts were flat—almost lifeless. And he made my skin crawl. Paul, do you remember how you used to feel when you came close to a snake? There's something wrong with that man."

"I know. I felt it, too. And it made the blood rush into my ears." Graham moved toward the communicator, placing his hand on the switch. "And you're right. I'll have to report immediately. They don't really need telepathy. And certainly, they never required real evidence. A suspicion is sufficient, and they'd be very suspicious if I didn't notify the sector leader tonight."

He depressed the switch deliberately, like a man firing a weapon. Then, he dialed a number, and waited.

The sphere lit, to show the face of Harold Bond.

"Oh, Graham." Bond frowned a little. "It's late. Do you have something to report?"

"Yes, sir." Graham's face was expressionless. "The mental communicator is finished. Do you wish to test it, sir?"

Bond opened his eyes a little more and nodded. "It's really done, then?"

"Yes, sir."

"I'll be there in a few minutes." The sphere darkened.

Graham looked at it. De-energized, the communicator seemed to be merely a large ball of clear material. It stood on its low pedestal, against its black background, reflecting a distorted picture of the chiaroscuro of the room. He leaned toward it, and saw a faint, deformed reflection of his own head and shoulders.

He spread his hands a little, and turned around. Elaine had crossed to the divan, where she sat, looking apathetically at the door, her hands folded in her lap. He smiled apprehensively, coughed, and held up a hand, two fingers crossed.

Elaine glanced at him, nodded, and resumed her watch of the door. Graham shrugged and walked over to his desk, where he stood, aimlessly looking down at the two headbands.

They both jumped convulsively when the buzzer sounded. Graham strode rapidly to the door, opened it, and stood back as the sector leader came in. Elaine had come to her feet, and stood rigidly, facing the door.

Sector Leader Bond closed the door, then looked from one of them to the other. He shook his head a little sadly, and waved a hand gently back and forth.

"Relax, you two," he said. "I'm alone this time." He turned to Graham. "Let's see what we've got."

Graham walked to his desk and picked up the two headbands.

"They're a little rough-looking, sir," he apologized. "But they work."

Bond tossed his head back with a little laugh. "They do look a little rugged, don't they?" he chuckled. "Well, we'll worry about appearance later. Right now, I'm curious. I want to see what these things do."

Graham handed over one of the bands and slowly adjusted the other to his head. For a moment, he looked searchingly at the sector leader, then his face relaxed into a relieved expression.

"Hear me?"

Bond had been examining the device in his hands. He looked up, puzzled.

"Of course I hear you," he said. "I'm not deaf."

Graham smiled a little, then placed a hand tightly over his mouth.

"Still get me?"

Bond cocked his head to one side, looked down at the device in his hands, then looked up again. "Well," he commented. "So that's the way they work. I thought you spoke."

Graham shook his head. *"Didn't have to. Try it on."*

Bond shrugged. "Well, here we go." He pulled off his cap, tossed it to a chair, and replaced it with the headband. For a moment, he looked around the apartment, then he glanced at Mrs. Graham. He blinked, ducked his head, and looked more closely at her.

"Ow! Nobody could be as bad as that!" He looked at Graham. *"What do you think?"*

"There's one outside." Graham inclined his head a little.

Elaine Graham sprang to her feet. "I'm terribly sorry," she apologized contritely. "It's just that I—"

Bond took off the headband abruptly. "I'm sorry, too," he said. "I was prying." He looked down at the device. "I'm not too sure about this thing," he added. "It works. I can see that much. But I'm almost afraid it works too well. What's it going to cause?"

Graham pulled off his own headband and extended his hand for the other. "I'm not sure," he admitted. "I'm not sure of anything at all." He frowned. "Wish I hadn't—" He looked at the sector leader quickly.

"I'm sorry, sir," he apologized. "Forgot my training, I guess."

Bond waved a hand. "Look," he said, "there are times, and there are places. Right now, I'm in your home, and I'm just as worried about this as you are. I'm just another person." He looked down at his neat uniform.

"Once," he mused, "we were all just people. Now—" He shrugged. "And then, these things come along." He looked at the two headbands, then at the man holding them.

"Wonder how many people feel like that?"

Graham held out the headbands. "I know one way to find out."

Bond nodded. "I see what you mean," he admitted. "But it could be pretty bad." He walked over to the chair and picked up his cap.

"Well," he added with a sigh, "I suppose I'd better grab these things and take them over to Research. Have to find out all we can about them. I've still got to report on them." Again, he looked at Graham. "You'd better come along, too. Research people might have a lot of questions, and I could never answer them."

Graham nodded and went to the hall closet. He took his coat from the hanger, put it on, and reached for his hat, then hesitated.

"You know," he said, "we might try one experiment, right here."

"Oh?" Bond raised his eyebrows.

"There's a man out in the parking lot. I believe he's detailed to keep watch on me. You might try him with one of the headbands. Then, see what he'll do with one on."

"Any special reason?"

Graham twisted his face uneasily. "I can't describe it," he said almost inaudibly. "You'd have to see for yourself."

Bond looked at him speculatively for a moment, then held out his cap and one of the headbands.

"Here, hold these."

He put the other headband on, accepted the first, and walked out of the apartment, followed by Graham, who still carried the cap.

As they came out and started across the parking lot, a man approached them.

Bond looked at him, frowned, then cast a sidelong look at Graham.

"That what you meant?" His thought carried an undercurrent of incredulity.

Graham nodded wordlessly, and Bond looked toward the approaching man again. Once more, his face wrinkled distastefully, then he spoke aloud.

"Oh, Ross. Want you to try some thing." He held out the headband he was carrying in his left hand.

Ross came up, accepted the device, and looked at it curiously. "You mean this is the thing he's been working on?" He jerked a thumb at Graham. "Saw his wife come out a while ago. Guess she had one of 'em on. She went right back in again."

Bond nodded. "This is it," he said. "Let's see how it works for you."

Ross shrugged. "Try anything once, I guess." He adjusted the band to his head, then stood, looking at the two men.

"Notice anything?" Bond looked at him sharply.

Again, Ross shrugged. "Nothing special," he said with a slight grunt. "Seems as though this guy's pretty nervous."

"You don't have to say anything, just think it. And see if you can communicate with Graham."

"Huh?" Ross had been looking directly at Bond. He frowned.

"You mean, this thing—" He paused, looking for a moment at Graham, then took the headband off. "Thing doesn't feel good," he complained. He held the device out to Bond, who accepted it.

"But it works? You could communicate both ways with it?"

"Oh, sure." Ross nodded grudgingly. "I got you, all right. But I couldn't get a thing out of this guy." He wagged his head toward Graham. "Except he was jittery about something."

"I see. Thanks." Bond accepted the headband. "We're going to take these to Research," he added. "Let the technicians there find out how good they are." He turned away and led Graham to his helicopter.

As Graham settled in the seat, he turned to the sector leader. "He just couldn't use it properly," he remarked. "Maybe only certain people *can* use them."

Bond nodded as he started the motor. "Or maybe only certain people can't." He busied himself in getting the machine up through the landing slot, then turned as they climbed into the night sky.

"Maybe you've got to be able to understand and like people before you can establish full contact with them. Maybe ... Maybe a lot of things." He was silent for a moment. "You know, this thing might become far more valuable than you thought, Graham."

Howard Morely looked up from a memo as the clerk tapped on the door.

"Come in."

The man opened the door and stepped inside.

"Sector Leader Bond is here, sir. He has some gentlemen with him."

"And what does he want?"

"He said it was about that new communicator, sir."

"Oh." Morely turned his attention back to the memo. "Have them wait." He waved a hand in dismissal and went on with his reading.

The beautification program was progressing well. Twenty miles of the old main highway through the valley had been completely cleared and planted. Crews were working on another stretch. The foreman of the wrecking crew down at the point, in Sector Nine, reported that the last bit of scrap had been removed from the old bridge support. Underwater crews had salvaged the cables and almost all of the metal from the fallen bridge itself, and the scrap was on the beach, ready for delivery to the reclamation mills in District One.

Morely smiled sourly. Harwood would have a storage problem on his hands in a day or so. The delay in delivery could be explained and justified. Morely had seen to that. Now, all the material was ready and could be delivered in one lot.

Harwood would have to raise his production quota in his community mills to use up the excess material, and that would slow down the clean-up in District One. The Old Man couldn't help but notice, and he'd see who was efficient in his region. The district leader pushed the memo sheets aside and placed his hands behind his head.

Slowly, he pivoted his chair, to look at the entertainment screen. He started to energize it, then drew his hand back.

So that crackpot, Graham, had finally come up with something definite. Morely smiled again. It had almost seemed as though the man had been stalling for a while. But the pressure and the veiled threats had been productive—again.

To be sure, the agents covering that project had reported that the device seemed to be merely another fairly good means of communication—nothing of any tremendous importance. But results had been obtained, and a communicator which was reasonably free from interception and which required relatively low power might be of some value to the community. He might be able to get a commendation out of it, at least.

And even if it were unsuitable for defense, there'd be a new product for one of the luxury products plants in the district, and the district would get royalties from the manufacturer. Too, it would keep people busy and make 'em spend more of their credits.

He grimaced at his vague reflection in the screen before him, and spoke aloud.

"That's the way to get things done. Make 'em know who's in charge. And let 'em know that no nonsense will be tolerated. Breathe down their necks a little. They'll produce." He cleared his throat and spun around, to punch the button on his desk.

The door opened and the clerk stood, respectfully awaiting orders.

"Send in Bond and the people with him."

The clerk stepped back, turning his head.

"You may go in now, sir." He disappeared around the door.

Harold Bond stepped through the doorway, followed by two men. Morely looked at them closely. Engineers, he thought.

"What have you got?" he demanded.

One of the men opened a briefcase and removed a large, dully gleaming band. Apparently, it was made of plastic, or some light alloy, for he handled it as though it weighed very little.

As the man laid it on the desk, Morely examined the object closely. It was large enough to go on a man's head, he saw. It had adjustable straps, which could be used to hold it in place, and there were a few spring-loaded contacts, which apparently were meant to rest against a wearer's forehead and temples.

A few tiny knobs protruded from one side of the band, and a short wire, terminated by a miniature plug, depended from the other.

The engineer dipped into his brief case again, to produce a small, flat case with a long wire leading from it. He put this by the headband, and connected the plugs.

"The band, sir," he explained, "is to be worn on the head." He pointed to the flat case. "To save weight in the band, we built a separate power unit. It can be carried in a pocket. We've tested the unit, sir, and it does provide a means of private communication with anyone within sight, or with a group of people. Two people, wearing the headbands, can communicate for considerable distances, regardless of obstacles."

"I see." Morely picked up the headband. "Do you have more than one of these?"

"Yes, sir. We made four of the prototypes and tested them thoroughly." Bond stepped forward. "I sent a report in on them yesterday."

"Yes, yes. I know." Morely waved impatiently. He examined the headband again. "And you say it provides communication?"

"Yes, sir."

"No chance of interception?"

Bond shook his head. "Well," he admitted, "if two people are in contact, and a third equipped person wishes to contact either one, he can join the conversation."

"So, it's easier to tap than a cable circuit, or even a security type radio circuit." Morely frowned. "Far from a secure means of communication."

"Well, sir, if anyone cuts in on a communication, both parties know it immediately."

Morely grunted and shook his head. "Still not secure," he growled. He looked at the papers on his desk. "Oh, put one on. We'll see how they work." He leaned back in his chair.

Bond turned to the man with the brief case, who held out another headband. The sector leader fitted it to his head, plugged in the power supply and looked around the room. Finally, he glanced at his superior. A shadow of uncertainty crossed his face, followed by a quickly suppressed expression of distaste.

Morely watched him. "Well?" he demanded impatiently, "I don't feel or see anything unusual."

"Of course not, sir," explained Bond smoothly. "You haven't put on the other headband yet."

"Oh? I thought you could establish communication with only one headset, so long as you were in the same room."

Bond smiled ingratiatingly. "Only sometimes, sir. Some people are more susceptible than others."

"I see." Morely looked again at the headband, then set it on his head. One of the engineers hurried forward to help him with the power pack, and he looked around the room, becoming conscious of slight sensations of outside thought. As he glanced at the engineers, he received faint impressions of anxious interest.

"Can you receive me, sir?"

Morely looked at Bond. The younger man was staring at him with an intense expression on his face. The district leader started to speak, then remembered and simply thought the words.

"Of course I can. Didn't you expect results?"

"Oh, certainly, sir. Do you want me to go outside for a further test?"

The headband was bothering Morely a little. Unwanted impressions seemed to be hovering about, uncomfortably outside the range of recognition. He took the device off and looked at it again.

"No," he said aloud. "It won't be necessary. It's obvious to me that this thing will never be any good for practical application in any community communications problem. It's too vague. But it'll make an interesting toy, I suppose. Some people might like it as a novelty, and it'll give them some incentive to do extra work in order to own one. That's what luxury items are for. And the district can use any royalty funds it may generate."

He laid the headband on his desk. "Go ahead and produce a few samples. Offer the designs to Graham's employer. He can offer them on the luxury market, if he wishes, and we'll see what they do. If people want them, it might be profitable, both for the district and for Consolidated." He shrugged.

"No telling what'll make people spend their credits." He started to nod a dismissal, then hesitated.

"Oh, yes. I think I'll keep this one," he added. "And you might leave a couple more. The regional director might be amused by them."

He accepted the two headbands and their power packs, put them in a desk drawer, and sat back to watch the three men leave the office.

After the door closed, he still sat, idly staring at the headband on his desk. He put it on his head again, then sat, looking about the room. There was no unusual effect, and he took the band off again, looked at it sourly, and laid it down.

Somehow, when Bond and those other two had been in the room, he had sensed a vague feeling of expectancy. Those three had seemed to be enthusiastic and hopeful about something, he was sure. But he failed to see what. This headband certainly showed him nothing.

He stared at the band for a while longer, then put it back on and punched the call button on his desk. As his clerk came into sight, he watched the man closely. There *was* a slight effect. He could sense a vague fear. And a little, gnawing hatred. But nothing was definite, and no details of thought came through. He shrugged.

Of course the man was fearful. He probably was reviewing his recent mistakes, wondering which one he might be called upon to explain. Too bad his mind wasn't clear enough to read. But what could you expect? Possibly, he could drive Research into improving the device later.

"Anyway," he told himself, "everyone has something they're afraid of. It's natural. And everyone has their pet hates, too." For an instant, he thought of Harwood.

He focused his mind on a single thought. *"Get me the quarters file for Sector Nine."*

There was a definite effect this time. There was a sharp radiation of pained surprise. Then, there was acquiescence. The clerk started to say something, then backed toward the door. The impression of fear intensified. Morely smiled sardonically. The thing was an amusing toy, at that. He might find uses for it.

He sat back, thinking. He could use it as a detector. Coupled with shrewd reasoning, well-directed questions, and his own accurate knowledge of human failings, it could tell him a great deal about his people and their activities.

For instance, a question about some suspicious circumstance would cause a twinge of fear from the erring person. And that could be detected and localized. Further questions would produce alternate feelings of relief and intensified fear. He nodded complacently. Very little had ever gotten by him, he thought. But from now on, no error would remain undiscovered or unpunished.

The clerk returned to place the file drawers convenient to his superior's desk. He hesitated a moment, his eyes on the headband, then picked up the completed papers from the desk and went out.

Morely riffled through the cards, idly checked a few against his notes, and leaned back again. The file section seemed to be operating smoothly. He looked at his desk. Everything that had to be done immediately was done. And the morning was hardly more than half over.

He rose to his feet. Surely, somewhere in the headquarters, there must be some sort of trouble spot. Somewhere, someone was not producing to the fullest possible. There must be some loose end. And he'd find it. He went out, jerking a thumb back at his office as he passed his clerk's desk.

"You can pick up those files again, Roberts. And see to it that my office gets cleaned up a little. I won't be back for a while."

He went out, to walk down the corridor to the snack bar.

There were a few girls there. He walked by their table, glancing at their badges. Communications people. He nodded to himself, ordered coffee, and chose a table.

As he glanced at the girls' table, he could detect a current of uneasiness. They'd probably been fooling away more time than they should. Too bad he couldn't get more definite information from their thoughts. Like to know just how long they had been there. He tilted his wrist, taking a long look at his watch. The current of uneasiness increased. No doubt to it, they'd been more than ten minutes already.

The girls hurriedly finished their coffee and left. Morely sipped at his own cup.

At last, he got up and went out. Might be a good idea to visit the Fixed Communications Section. Looked as though there might be a little laxity there.

As he walked down the corridor, he mentally reviewed the operation of communications. There was Fixed Communications, responsible for communicator service to all the offices and quarters in the district, as well as to the various commercial organizations. There were also Mobile Comm, Warning, Long Lines, and Administrative Radio.

Of these, the largest was Fixed Communications, with its dial equipment, its banks of video amplifiers, the network of cables, and the substation equipment. It would take days to thoroughly check all their activities. But the office was the key to the entire operation. He could check their records, and get a clue to their efficiency. And he could question the section chief.

He took the elevator to the communications level and walked slowly along the hallway, glancing at the heavy steel door leading to Warning as he passed it. That could be checked later, though there would be little point to it.

It had always annoyed him to think of the operators in that section. They simply sat around, doing nothing but watch their screens and keep their few, piddling records. They did nothing productive, but they had to be retained. Actually, he had to admit, they were a necessity under present conditions. War was always a possibility and the enemy was building up his potential. He might strike at any time, and he'd certainly not send advance notification. If he did strike, the warning teams would perform their brief mission, alerting the active, working members of the defense groups. Then, they would be available for defense. And the defense coördinators required warning teams and equipment in prescribed districts. His was one of these.

He grumbled to himself. Even the number of operators and their organization were prescribed. This was a section, right within his own district, where he had little authority. And it was irritating. Drones, that's what they were.

He continued to the Fixed Communications office. Here, at least, he had authority.

He walked through the door, casting a quick glance at the office as he entered. The section chief got up from his chair, and came forward. Morely felt a little glow of satisfaction as he detected the now familiar aura of uneasiness. Again, he wished this device he wore were more effective. He would like to know the details of this man's thoughts.

"Good morning, sir." The Fixed Communication chief saluted.

Morely returned the salute perfunctorily, then examined the man critically.

"Morning," he acknowledged. "Kirk, I want you to get some new uniforms. You look like a rag bag."

A little anger was added to the uneasiness. Kirk looked down at his clothing. It wasn't new, but there was actually little wrong, other than the slight smudge on a trouser leg, and a few, small spots of dullness on his highly polished boots.

"I've been inspecting some cable vaults, sir," he explained. "We had a little trouble, due to ground seepage."

"It makes no difference," the district leader snorted, "what you've been doing. A man in your position should be properly attired at all times." He paused, looking Kirk over minutely. "If your cable vaults are in such bad condition, get them cleaned up. When I look your installations over, I shall expect them to be clean. Clean, and in order."

He looked beyond Kirk. "And get that desk cleared. A competent man works on one thing at a time and keeps his work in order. A place for everything, and everything in its place, you know. You don't need all that clutter. Is the rest of your office as disorderly as this?"

He looked disparagingly about the small room, then turned toward the door to the main communications office. Kirk moved to open the door.

At one side of the large office was a battery of file cabinets. Four desks were arranged conveniently to them. Morely looked at this arrangement.

"What's this?"

"Billing and Directory, sir. These are the master files of all fixed communication subscribers. From them, we make up the semiannual directory, its corrective supplements, and the monthly bills."

Morely frowned at the desks and files, then looked at the clerks, who were bent over their desks. As one of the girls straightened momentarily, he recognized her. He'd seen her earlier, in the snack bar. He looked more closely at her desk. She had reason, he thought for that radiation of uncertain fear he could sense.

"What's in those files?" he demanded.

"It's a complete index to all subscribers, sir." Kirk looked a little surprised. Morely recognized that the man thought the question a little foolish. He cleared his throat growlingly.

"Let's see one of those cards."

Kirk walked to the file, pulled a small envelope at random, and held it out. The district leader examined it.

"Hah!" he snorted. "I thought so. Duplication of effort. This has nothing on it that isn't in my quarters and locator files."

"There's billing information on the back, sir," Kirk, pointed out. "And current charge slips are kept in the envelope. We use these to prepare the subscriber bills, as well as to maintain the directory service. It's a convenience file, to speed up our work."

Morely turned the envelope over in his hands. "Oh, yes." He opened the envelope, to look at the slips inside. "How do you get the information for these?"

"The charge slips come from Long Lines, sir." Kirk paused. "We get billing information for basic billing from the counters in the dial machine. The other information comes from installation reports and from the quarters file section and the locator files."

Morely handed the envelope back.

"I can see, Kirk," he said, "that you've built up a whole subsection of unnecessary people here." He stepped over to the file cabinets, examined their indices, then pulled a drawer open. He pulled his notebook out, consulted its entries, and searched out an envelope. For a moment, he compared it with the notebook. Then, he turned, holding out the envelope.

"And you don't even keep your information current," he accused. "This man was transferred yesterday afternoon, to another sector. You still show him at his old quarters, with his old communicator code."

"We haven't that information from Files yet, sir," protested Kirk. "They send us a consolidated list of changes daily, but it generally doesn't come in till thirteen hundred."

Morely dropped the envelope on one of the desks.

"Quarters Files can handle this entire operation," he declared, "with a little help from Fiscal. And they can handle it far better than your people here." He stopped for a moment, thinking, then continued. "Certainly," he decided, "Fiscal can take care of your billing. They handle the funds anyway, in the final analysis. And you can coördinate your directory work with the chief clerk at Files. You've got excess people here, Kirk. We don't need any of them."

He looked at the desks and felt a wave of consternation. Kirk spread his hands.

"But we have the information we need close at hand, sir. Our directory has been coming out on time, and in accurate condition. And our billing is well organized. The directory and billing are my respons—"

Morely waved a hand, then tapped himself on the chest with a long forefinger. "The entire operation of this headquarters is my responsibility, Kirk," he said positively, "and mine alone. And I mean to take care of it. You're responsible to me that Fixed Communications are kept in order, and I don't mean to relieve you of a bit of that responsibility. But I won't have you making jobs and wasting funds on excess personnel." He snorted. "Convenience files are all right. But they're meant to save work, not make it."

Kirk shook his head. "A decentralization will make it difficult," he began.

Again, Morely cut him off. "Don't start telling me why you can't do something," he snapped. "Work out a way you *can* do it. Make up plans for transferring this filing function to Quarters Files, and work up a plan for transferring your billing to Fiscal. That's their business, and they know how to handle it. Submit your study to me this afternoon." He looked around the office again.

"The people in Files and Fiscal can handle this workload without adding a single person. And they will. You're using four clerks to swing it. Kirk, I want this organization to run efficiently, and excess personnel don't lead to economic operation." He stared at the section chief.

"Give these four people their notices today, and I'll expect some suggestions from you as to further streamlining of your section within the next two days. And be sure they're sound suggestions, which result in personnel savings. Otherwise, I'll be looking for a new section chief up here."

For a few seconds, he stood, enjoying the waves of consternation and futile anger which beat about him. Almost, he could pick up some of the despairing thoughts in detail. The clerks, of course, were second-class citizens. And without employment, they'd soon lose their luxury privileges. Unless they were fortunate enough to find other employment very soon, they'd have to move to subsistence quarters, and learn to do without all but the most meagre of food, clothing, and shelter. When they did get employment again, they'd appreciate it. He looked majestically around the office once more, then turned and strode away.

He went through the corridor to the elevator, and stepped in, smiling contentedly. The morning hadn't been entirely wasted.

As he got out of the elevator on executive level, he glanced at his watch. It wasn't quite time for lunch, but there would be little point in spending the few remaining minutes in his office. He walked slowly toward the executive cafeteria.

After lunch, he returned to his office. A few matters awaited his examination and decision, and he busied himself for a short time, disposing of them. He paused over the last.

It was a request from Kirk for more cable construction. The justification showed figures which indicated an increase in executive type communications during the past few months. This, coupled with new quarters construction, necessitated additions to the cable trunks from the main exchange. There was added a short survey of necessary repair to existing cable facilities.

Morely leaned back. If he approved the request, he would be helping Kirk increase his section. On the other hand, if he disapproved it, and the communicator lines became congested, he might find himself open to criticism later. Some of his satisfaction evaporated. He looked sourly at the paper.

Suddenly, he thought of Bond's new project. The man had claimed this device could serve as a communication means between its wearers, and had demonstrated that his claim had some truth. After noting the slight fatigue the device seemed to cause in this application, and the vagueness of the device's operation, Morely had disregarded the claim. But junior executives could put up with a little fatigue and inconvenience. And he could see that they did. It might even cut down the time they were always wasting, talking with one another. He rubbed his chin with one hand.

"Well," he told himself, "let's see how it works."

From the way Bond had acted in his office, the sector leader might be still wearing his headband. In fact, he probably was. Morely concentrated on the man, then concentrated on a single, peremptory thought.

"Bond! Can you receive me?"

The answer was prompt. *"Yes, sir. You wanted me?"*

"Of course, Idiot. Why do you think I called? Do you really believe these things would be suitable for routine communication? Could they supplement our normal system?"

"Certainly, sir. They should be very effective."

"Have you offered them to Consolidated yet?"

"Yes, sir. They've accepted them. They're beginning to tool up for production."

Morely winced. He had given the order, to be sure—and before creditable witnesses. Bond had been right in taking immediate action, and his speed would have been commendable in most cases. But this time, Morely regretted his subordinate's efficiency. It was possible the devices might have a practical use after all. Possibly he had been hasty in releasing them to the open market. He shrugged away his thoughts. After all, an administrator had to make quick decisions. He returned to his unusual conversation.

"Set up a line in research and make up sufficient of those communicators to outfit the executive personnel of this district."

"Yes, sir."

"And give me delivery as soon as you possibly can. How soon will that be?"

"We can do it in five days, sir."

"Make it three. That's all."

Morely took off his headband. It wasn't as good as a communicator sphere, but it would be good enough. He looked at the request from Communications. Possibly, he would be able to cut Kirk down still more. He scrawled a "disapproved" on the sheet and initialed it. He started to toss the sheet to the corner of his desk, then hesitated.

Drawing the request back to him, he added: "Two subjects on same request. Resubmit as separate requests." He tossed the sheet to the desk corner, for the clerk to pick up. Let Kirk make up new requests, then worry about why his new construction request was still disapproved. He could always be advised to resubmit later, if the headbands didn't work out.

Miles away, Bond turned to an engineer.

"Tool up and start producing these communicators as fast as you can make 'em, Morris. I'll tell you when to stop. The Old Man just ordered a batch of 'em, and this is one order I want to comply with, and fast!"

He walked toward the small production office. Let's see, he had to produce enough for all the exec personnel in the district. Have to start finding out just how many of those guys there were.

"Make delivery as soon as possible, huh? Cut my estimate by two days? I'll have 'em out over night, if I have to start driving people to do it."

Morely looked up as the communicator beeped. He reached to the control panel and touched the switch. The face of his deputy appeared in the sphere.

"The section chiefs and field leaders are in the conference room, sir."

"Very good." Morely pushed back his chair. "I'll be right in."

He stepped through the door and crossed the outer office to the conference room. As he entered, there was a rustle of motion. The section chiefs and field leaders stood at attention around the table, waiting. At each place at the table was a blank notepad. The district leader went immediately to the head of the table and sat down.

"Gentlemen," he began, "I'll make this short. I've called you in to try out a new device which I intend to use to help solve the ever-present problem of communication." He looked toward Ward Kirk, who had glanced up in surprise.

"From time to time," he continued, "requests for more and more communicator lines have been coming in to my office. Since no one else seemed to be able to do anything about it, I decided it was time for me to step in. After all, we can't expand our cables indefinitely. We haven't unlimited funds at our disposal and there are other projects demanding attention. Important projects.

"A new electronic development has come to my attention, and it promises to relieve the load on our communicators. Each of you will be issued one of these devices, which I believe are called 'mental communicators,' or something of the sort. And you will draw sufficient of them to outfit those of your people who have occasion to use communication to any large degree. You will use them for all routine communications." He nodded to his deputy, who stepped to the door and beckoned.

Two men came in, carrying cartons, which they distributed around the room. Morely waited until one of the cartons was in the hands of each of the men before him, then he reached up to touch the headband he was wearing.

"This is the device I'm speaking of," he said. "Each of you will wear one of these at all times while you are on duty. You will find, after a little practice, that you will be able to call any associate who is similarly equipped. And you will use them in place of the conventional communications whenever possible." He cleared his throat raspingly.

"Sufficient of these devices have been produced to outfit all the key people of this district. I shall leave it to you to distribute them to your subordinates, and to instruct those subordinates in their use. And I shall expect the load on our communicator cables to be appreciably diminished." He looked to one side of the room.

"Bond."

"Yes, sir."

"You will instruct those present in the use of this new communicator." Morely rose and left the room.

As the district leader disappeared through the door, Harold Bond walked to the front of the room. In his hands, he held one of the headbands and a power pack.

"Gentlemen," he said, "this is a form of communicator. I don't pretend to understand precisely how it operates, though I watched its development and set up a production line for it. All I know is that it works. And I know how to use it — to some extent.

"The district leader remarked that one could learn to use it with a little practice, and he's right. Basically, anyone can use it as soon as he puts it on for the first time. But it's like so many other tools. The more you use it, the more proficient you get with it. And I suspect it has capabilities I haven't found yet." He shrugged.

"Operation is simple in the extreme. Since the first model, refinements have been added, and it's unnecessary now for an operator to make any adjustments, other than intensity."

He picked up the power pack.

"This is the power pack, which is plugged into the headband, thus." He paused as he connected the two plugs.

"If you gentlemen will perform the operations as I do, this will take only a short time."

There was a crackling in the room as cartons were opened. Power packs and headbands rattled against the table for a moment, then Bond continued.

"Having plugged in the power pack, you turn this small knob very slightly in a clockwise direction, then place the headband on your head. The knob is the switch and intensity control, and it's quite sensitive. Most people need very little intensity. If you have difficulty with communication, raise the intensity a little at a time, till thoughts come through clearly." He paused, as the men before him adjusted the headbands to their heads.

"The power pack," he continued, "may be placed in a pocket." He reached down. "Personally, I carry mine in my shirt, since I find that convenient."

He looked around the room. Men were turning to stare at their neighbors. Bond could detect a current of uncertainty, then a sensation of pleased surprise. Snatches of thought drifted to him. He ignored them for the moment. Time enough to become acquainted with people later. He placed a hand over his mouth, so everyone could see he was not speaking.

"Can everyone receive me?"

There was a wave of affirmation, and Bond nodded.

"Simple, isn't it? Are there any questions?"

A jumble of thoughts made him waver. Most of them could have been phrased, "How does this thing work? What does it do? Am I dreaming?" Bond smiled in real amusement. He held up a hand.

"I felt the same way," he thought reassuringly. *"Sometimes. I still do. All I can tell you is what you've already found out for yourselves. It works. I'm told it's*

a sort of telepathic amplifier and radiator. But as I told you, I don't understand its principles. As to practice? I'm still meeting interesting people. So will you." He took off the headband.

"If anyone has any further questions on operation, I'll try to answer them," he thought quickly. He glanced around the room. Three men were looking at him blankly. He took careful note of them, and mentally shook hands with himself. They were the ones he'd thought would blank out. He spoke aloud.

"I'm sorry, gentlemen," he apologized. "I forgot I might be out of communication. I'm not completely used to this mentacom, myself." He looked toward the deputy leader.

"Do you have anything to add, sir?"

The deputy shook his head. "No," he said thoughtfully. "I think the demonstration was adequate. He cast a quizzical look at Bond, then looked around the room.

"You gentlemen will find a supply of these devices in the outer office. You may draw one for each person you wish outfitted. If any of you have further questions, I would suggest you get in touch with Community Research. They understand this thing." He waved toward the door. "This meeting is adjourned."

He watched as the men filed from the room, then turned on Bond.

"What was that business after you took off your headband?" he demanded. "I received you perfectly, and so did practically everyone here. Why the apology?"

Bond grimaced. "We found out something peculiar while we were making preliminary tests on this device, sir," he explained. "Some people don't seem to be able to pick up clear thoughts with it, unless another person uses the mentacom to drive in to them. Most of us can pick up thoughts from anyone we look at, whether they have a band on or not. Definite, surface thoughts, that is."

"And?" The deputy's expression was still questioning. He reached up to point at the band he was still wearing. "I'm getting some mighty peculiar secondary thoughts right now," he added.

"And the people who can't use the device fully have other peculiarities, sir. I'd rather not go into detail. You can find out the whole story for yourself with a very short bit of experimentation, and you have a subject right at hand. If I simply told you, you probably wouldn't believe me anyway."

The deputy nodded slowly. "For the moment," he said, "I'll take your words—and your thoughts—as true. Now, one more question: Can a person, using one of these things, successfully lie to another person who wears one?"

"No, sir." Bond was positive. "It's impossible."

"I got that impression. Thanks." The deputy turned and walked out of the door. Bond looked after him, a slight smile growing on his lips.

"Old Man wanted 'em," he told himself. "He's got 'em."

The Fiscal chief glanced through the letter in his hands, then canted his head a little and read again. He lowered it to his desk, then sat for a moment, to stare into space. Finally, he looked down once more.

Central Coördination Agency
Office of the Comptroller

CCA 7.338   21 July, 2012

To:	District Leader
	District Twelve
	Region Nine
Attn.:	Fiscal Chief
Subject:	Mental Communicator

1. It has been brought to the attention of this office that a product known as the "Consolidated Mental Communicator" is being manufactured in District Twelve, Region Nine, and offered for sale as a luxury item.

2. The characteristics of this device have been investigated by the Technical Division, Central Coördination Agency, and it has been found that the device does in fact permit communication between persons by telepathic or some similar means.

3. This device is presently being offered for sale in retail luxury stores throughout the nation. The volume of sales and of potential sales warrants distribution of the manufacturing load to manufacturers other than the Consolidated Electronics Company, who, it is understood, presently hold an exclusive manufacturing agreement with the office of the District Leader, District Twelve, Region Nine. This arrangement is inconsistent with the sales and use potential of the device in question.

4. The agreement between District Twelve, Region Nine, and the Consolidated Electronics Company will be forwarded

immediately to this headquarters for consideration. It is contemplated that this agreement will be terminated and replaced by a manufacturing license from the Products Division, Central Coördinating Agency, who will further license other manufacturers to produce this device.

By Command of Chief Coördinator Gorman

KELLER
Comptroller

MRK/pem

The Fiscal chief shook his head. This one spelled trouble—in capitals. The royalty payments from Consolidated had become one of the major sources of income for the district. And Morely had ordered project after project, using those funds to pay for them. Some of the projects were still outstanding. The Old Man would blow his top.

He looked again at the small scrap of paper which was clipped to the letter. On it was scrawled: "DeVore—See me—HRM."

For a moment, DeVore considered using his own mentacom, then he discarded the idea. To be sure, the leader had insisted that his subordinates use the devices for their own communications, and he'd cut Fixed Communications to the bone. But he still insisted on either communicator calls or personal contact when he wished to talk to any of his people. And he discouraged any but essential use of the communicator system, generally demanding that people come in to see him.

DeVore wrinkled his face disgustedly. It *was* hard to communicate with the district leader by means of a headband. There was a repellent characteristic about the man's mental emanations, and he seemed to fail to comprehend nuances of meaning. Similes, he ignored completely. Thoughts had to be completely and clearly detailed, then phrased into normal, basic wordage before he would acknowledge them. None of the short-cuts used by other members of the administrative staff seemed to work out in his case. He apparently didn't notice visualizations, and he never made one. His transmission was as stiff and labored as the type of communication he required from others—more so, if anything. DeVore scratched his neck.

"How," he asked himself, "does one define a telepathic monotone?"

There were a few others with whom DeVore had experienced similar difficulties, but most people, he had found, picked up meanings and concepts without difficulty—even seemed to anticipate at times. And since the new induction mentacoms had come on the market, with the annoying contacts and headstraps removed, virtually everyone seemed to be either in possession of one of the devices, or about to get one. And, they were worn everywhere.

He smiled as he thought of the young father-to-be, who had bored through the evening traffic rush yesterday. The youngster had been so intent on getting his wife to the hospital that he'd probably failed to see half the ships that clawed out of his way. And his visualization had been almost painfully clear. He'd probably be apologizing for weeks to everyone he contacted.

DeVore straightened in his chair. What would happen, he wondered, if the leader ever ran into one of those situations?

"Yipe!" he muttered. "What a row that would be."

He shrugged, got out of his chair, and walked out into the corridor.

"Better get it over with," he told himself.

As he approached the leader's door, it opened, and Ward Kirk came out. He closed the door with a careful gentleness, then faced it for an instant. DeVore was conscious of a wave of hopeless fury, and a fleeting glimpse of Morely's face, framed by brilliant flame. Then, Kirk faced around and saw him.

"Careful," DeVore thought. *"You're broadcasting. He'll pick you up."*

Kirk grimaced and DeVore saw a faint image of a tyrannosaur, which reared up, jaws agape. Blood dripped from the human figure gripped in the creature's talons.

"The old ... wouldn't understand if he did."

DeVore grinned. *"See what you mean. Well, guess I'm the next victim."*

He stepped to the door and tapped.

"Come in."

Morely looked up as his Fiscal Chief entered, then swept some papers aside. "Well, what do *you* want?"

DeVore held out the letter. "You wanted to see me, sir, about this." He placed the paper within the reach of his superior, who snatched at it, held it up for a moment, then dropped it to his desk.

"Yes, I did. What can we do about it?"

"Why," DeVore spread his hands slightly, "we'll have to comply."

"That isn't what I meant, Idiot! How can we continue to receive the payments from Consolidated?"

"I don't think we can, sir. If Central Coördinating wants to put the device on a national basis, we can't do anything about it."

Morely looked down at the letter, then glared searchingly at DeVore. "The way I read this," he declared, "they want to distribute manufacturing rights on the communicator to plants in other regions than this. Right?"

"Yes, sir."

"But they don't say anything about our continuing the Consolidated payments on an overwrite basis, for the sale of devices they may make. Now, do they?"

"No, sir. But that's implied. In cases like this, Central always takes over all rights." DeVore hesitated. "I believe regulations—"

"I don't care what's implied, DeVore. And I don't care what you believe. All I see is what's in this letter. They want to distribute the manufacturing load, and I'm quite willing that they should. I want to continue receiving the payments from Consolidated. Now, you arrange it so that they're satisfied and I'm satisfied."

"But that'll mean Consolidated will have to pay double. We can't—"

"Don't say 'can't' to me!" Morely held up a hand angrily. "DeVore, I'm not going to tell you how to do this. I want it done. The details are your affair, and if I have to teach you your business, I'll get someone who can do things without having to have them spelled out to him." He leaned back, to glare at DeVore.

"Now, get on the job. I told you to make arrangements for me so that we will retain our payments from Consolidated. And I'm not interested in what arrangements you make with them, or what arrangements they make with Central. Is that a simple enough order for you to understand?"

"Yes, sir. I understand all right. But—"

"Good! I'm glad I managed to get at least one simple idea into your head." The spring in the chair twanged as Morely came forward, to poke his head at DeVore. "Now, get to work on it."

He jerked his head down for a quick look at the letter on his desk, then looked up again.

"And I'll expect a report from you by tonight that you've got the matter taken care of."

DeVore looked at his superior expressionless for a heartbeat. He had been given peculiar orders before, and he'd always managed to work out the problems involved. But this was the ultimate. This one seemed to be just plain illegal. And there was no point in arguing further. There was just the barest chance that there might be some legitimate way out. If he challenged the Old Man on an illegal order, he just might get his ears pinned back. He'd simply have to go back to his office and try to hunt out a technicality. He nodded.

"Yes, sir. I'll get on it immediately."

He saluted and started to leave the office. But he didn't make it.

"And, DeVore!"

The Fiscal chief halted abruptly, and turned.

"Sir?"

"I'm getting tired of the negative thinking you people seem to have fallen into lately. I'm sick of going into every routine detail with you. When you got that letter, you should have immediately worked out a method of retaining the royalties. Then, you could have come in and presented it for my approval. That is the kind of work I want. And that's the kind of work I mean to get in the future. Do you understand?"

Sternly, DeVore suppressed a sarcastic thought. He held his mind and face blank and nodded with a semblance of respect.

"Yes, sir."

"Very well." Morely waved a hand. "Now get something done."

As DeVore walked through the corridor, he thought over the situation. Of course, the easy way out would be to force Consolidated to continue the payments in addition to their license fees from Central. That could be done. There were all kinds of methods by which pressure could be brought to bear on any company by the district leader's office. And from Consolidated's point of view, double payments could offer a cheap means of keeping out of difficulties. They would be able to pass most of the cost to the consumer by a slight price increase, justified by a minor modification of the devices.

But they wouldn't be happy about it, and there would come a day when an auditing team from Central would be checking in the district. And that would be the day of days!

DeVore turned in at the door to his own office, crossed the room, and sat down at his desk.

To be sure, he could request a share of the fees from Central, and they'd make an award. But they'd never award more than fifty per cent, and it'd be hard to get that much. That was no good. The Old Man would want the same payments he'd been getting.

Or, he could try to negotiate a new agreement with Consolidated, double the royalties, and then request fifty per cent from Central. He grinned wryly. That would be within legal limits, he was sure, but Central knew the present arrangement, and he knew that they knew. And so would most of the interested manufacturers in other regions. The first-class citizens who owned the plants had their own liaison. They'd all balk. Then, Central would invalidate both old and new agreements and refuse compensation of any kind to district. That would be a suicidal course.

He looked up, thinking of one of the girls out in the legal crew.

"Fiscal regulations, please. And Markowitz on royalties, too."

The girl turned half around, and he could see a faint impression of her view of office details. Then, she went to a book rack. For a few seconds, she glanced over the books, then selected two large volumes.

"Shall I look it up, or do you want the books?"

"I'll take them. Might need quite a bit of research."

Shortly, the girl appeared in his doorway. Quickly, she laid the two volumes on his desk.

DeVore nodded his thanks and opened regulations. Some of the paragraphs were delightfully vague, and could be subject to more than one interpretation. But one paragraph was clear and explicit. And that was the one he was concerned with.

A royalty agreement with, or manufacturing license from Central Coördination definitely abrogated any agreement with, or payment to, any lesser headquarters. Such an agreement or license barred any further negotiation between any lesser headquarters and a manufacturer, relating to the product concerned. Double royalties were prohibited in any case.

He pushed the books aside. There was no need of looking in Markowitz. That regulation paragraph took care of this exact situation, and disposed of it neatly. For an instant, he thought of taking the volume in to the leader's office. Then, he remembered the threatening note in the authoritative voice and the flat, deadly thoughts he had noted as secondaries.

That wouldn't work either. He thought of the undercurrent in Kirk's thoughts. Kirk had been carrying a regulation book, he remembered. He contacted the Fixed Communications chief.

"Don't," he was told. *"I tried it. Know what happened?"*

"Go ahead."

"He got the regional director on the communicator. I've been transferred to Outpost. They seem to need a cable maintenance chief up there. And I was lucky at that. I started to protest, and they nearly had me for insubordination." Abruptly, Kirk cut away.

DeVore stared unseeingly across the desk. He'd been at Outpost for a short time once, on an inspection trip, and he still remembered the place. At one time, it had been a well supplied, well organized post. At that time, observational duty had been regarded more highly than now, and the place had been desirable for any single officer, though the married men had objected to being separated from their families by the many miles of frozen waste. But that had changed.

Now, Outpost was the end of the line. The dilapidated surface quarters offered poor protection from the fierce cold. Supply ships were rarely scheduled to the place, and were often held up by storms when they were scheduled. Half rations—even quarter rations—were commonplace. He shook his head. Kirk was in real trouble, and there would be no point in joining him. That would help neither of them.

This, he thought, was a situation. Then, he realized something else. From Morely's point of view, it was a perfectly safe situation, with nothing to lose. The district leader could easily disclaim any responsibility for his Fiscal chief's actions in this matter. After all, he hadn't given any detailed instructions. He had made no direct suggestion of any illegal course. He'd merely consulted his Fiscal expert on a technical matter, and if DeVore had seen fit to use an illegal method of solving a problem, it was DeVore's responsibility alone.

To be sure, Morely had been a little emphatic in his order, but that was simply because he was well aware of his Fiscal chief's disinclination to make exhaustive technical research.

DeVore pursed his lips and looked thoughtfully at the regulation book. He might be able to use the same tactic Morely was following—if he were so inclined. He could issue verbal instructions to the sector leader concerned, and Bond might fail to see the trap. Then, he could report to the leader that the matter was taken care of, indorse the letter back to Central, with the agreement copy, and let Bond turn in funds under one of the "miscellaneous received" accounts. In fact, he realized, that was just about what the district leader expected him to do.

He smiled and shook his head. A few months ago, it was possible he could have done that, but even then, he wouldn't have. And now, with the mental communicators in use, it would be a flat impossibility. The trap would be as obvious to Bond as it had been to him. He leaned back in his chair and tapped his fingertips against each other.

The mentacoms, he knew, were in common use by this time, in virtually every office of district, regional, and national administration, as well as by most citizens. And he'd served under Marko Keller once—known him fairly well, too. He shrugged.

It would be a little irregular for a district Fiscal chief to make direct contact with the Coördination Agency's comptroller, but there was nothing like getting the most expert and authoritative advice available. He relaxed, trying to recreate his memories of the man who was now National Comptroller.

Marko Keller strode purposefully into the filing section. He could easily get the data he needed by simply contacting one of the clerks, he knew, but he felt an urgent need for personal activity. That conversation with DeVore, way out in Region Nine, had upset him more than he liked to admit, even to himself.

It wouldn't be so bad if it were an isolated incident. Such things could be taken care of by administrative action, and a single instance would cause little disturbance. But there were too many, happening too often. He pulled a file drawer open, violently.

One of the clerks approached. "Can I help, sir?"

Keller turned to look at him. The man, he noted, was wearing one of the late model inductive headbands that had been sold in such quantities lately. Deluxe model, too. Must have cost him at least two months' pay. Like almost everyone else, he was vitally concerned in this latest affair. Keller frowned. He, himself, he realized, was acting childishly. He would simply be wasting time by trying to do this by himself.

"Yes," he growled. "Get me a brief on a few cases like this one." He made full contact with the man, rapidly summarizing his conversation with DeVore, and including DeVore's short flash of his own conversation with Ward Kirk.

"And get a rundown from personnel. Dig up something on their angle, too. Several representative cases. Get a few people to help you—many as you need. I'm going to take this whole mess in to the Chief tomorrow morning."

Paul Graham swept into the apartment, seized his wife about the waist and swung her into the air, to set her on top of one of his bookcases.

"They've done it, honey," he shouted.

Elaine kicked her heels in a rapid tattoo against the back of the case.

"Paul Graham, you get me down this instant," she ordered indignantly. "Who's done what?"

Graham stepped back and beat on his chest. "Meet the new production manager, Mentacom Division, Consolidated Electronics."

"Production manager? But, Paul, only first-class citizens can hold supervisory positions."

"Not any more. Didn't you have the communicator on for the news? It all came in."

Elaine shook her head and jumped to the floor. "I've a confession to make, Paul. Ever since they stopped the compulsory notices, I haven't had the thing on at all. It bothered me."

Her husband shook his head in mock dismay. "So now, I'm married to an ignoramus." He spread his hands. "She doesn't know what's going on in the great, big world." He shook a finger at her.

"It all busted this afternoon, darling. While you sat around in your splendid isolation, everything turned upside down."

She looked at him indignantly for an instant, then turned toward the kitchen.

"Paul, if you don't stop raving, I'm going to get my mentacom and pry it out of you," she threatened. "Now, you just settle down. Stop talking in circles and tell me what this is all about."

"Oh, all right. If you insist." Graham sank into a chair, looking like a small boy caught in a prank. "First, there are no more first-class citizens— no second-class citizens—not even third-class citizens. Everyone's a citizen again. Period." He threw his hands up.

"You mean—?"

"That's exactly what I mean. No more restrictions. No more compulsory community work. No more quarters inspections. And no more privileges. We've got rights again!

"If you want a dress, you buy it. You don't worry about whether it suits your station. If I can hold a job, I get it. And I did!" He got out of the chair and strode across the room, to sit on the arm of the divan. "And I can do this, if I want to. If I break this thing down, so help me, George, I'll go out and buy a new one." He bounced up and down a little.

"The administrators are going back to their original jobs. They're responsible for defense, in case of enemy attack, and that's all." He paused. "Of course, until sector and district elections can be held, they'll still take care of some of the community functions—some of them, that is. But the elections'll be set up in a few weeks, and we'll be able to choose our own officials for community government."

He bounced to his feet again, strode around the bookcases, and looked down at his desk. Then, he looked around again.

"Corporations are being set up to take over home construction." He held up a hand. "*Home* construction, I said, not quarters. They're commercializing helicopter manufacture, all kinds of repair work, and a lot of other services. And they're going to restore patent rights. That means plenty to us, darling, believe me."

"But, but why? What happened?"

Graham turned on her. "Elaine," he cried, "haven't you noticed how many people are wearing mentacoms now, all the time? Haven't you noticed the consideration people have been giving each other for the past weeks? Remember what I told you once? If you fully understand a person, you simply can't kick him around. It's too much like taking slaps at yourself. With the exception of a few empathic cripples, who can't use the mentacom properly anyway, everyone, inside the administrative offices, as well as out, recognized that the bureaucracy was simply unworkable as it stood. So, they changed it. Effective immediately."

Elaine stamped her foot. "You know I haven't been out of this apartment," she cried. "And you know why. I simply couldn't stand the treatment I got. I'd have gotten into serious trouble in minutes. So, I've stayed in. I've done my shopping by communicator, and contented myself right here." She paused.

"But how is the new administration going to be supported? What are people going to do? How are they taking it? It's all so sudden, I should think—"

Graham held up a hand.

"Hey," he protested. "One at a time, please! First—remember taxes? Remember how we used to growl about them? They're back. And I love 'em. Second—nobody is going to do anything. Anything drastic or unusual, that is. And finally? Everyone I've seen is taking it in their stride. Seems as though they've been sort of expecting it, ever since they started mind-to-mind communication.

"You'd be surprised how good most people are at it, now that they're used to it. You start into a line of helicopters. All at once, you realize that the guy coming is really in a hurry. He's got to get somewhere, fast. So, you let him go by. The next fellow's not going to be in any tearing rush. He'll let you in, and cheer you on your way.

"You feel like being left alone? Nobody'll even notice you. But if you feel like talking, half a dozen total strangers'll find something in common with you. And they'll discuss it. Honey, you'll be surprised how much you've missed. Get your mentacom. Let's take a little shopping trip."

"And here's one of our more difficult cases. But he's coming along nicely." Dr. Moran pointed through the one-way window.

"Name's Howard Morely. He used to be a district leader, under the bureaucracy. But along in the last few weeks, just before the change, he got into some sort of scrape. They questioned him, and declared him unfit for service. Put him out on a pension." He pulled at an ear.

"Matter of fact, I understand his case had quite a deal to do with the change—sort of triggered it. They tell me it sort of pointed up the fallacies of the bureaucracy." He shrugged.

"But that's unimportant now, I guess. He almost receded into complete paranoia. Had a virtually complete case of empathic paralysis when he came to us. Simply no conception of any other person's point of view, and a hatred of people that was fantastic. But he's nearly normal now."

The visiting psychiatrist nodded. "I've seen the type, of course. We have a number of them, too. You say this new technique was successfully used in his case?"

"Yes. We had doubts of it, too. Seemed too simple. Sure, we're all familiar with the mentacoms by now. Wouldn't be without my own. But the idea of a field generator so powerful as to force clear impressions into a crippled mind like his, without completely destroying that mind, seemed a little fantastic." He shrugged.

"In this case, though, it was a last resort, so we tried it. He resisted the field for days. Simply sat in his cell and stared at the walls. We were almost ready to give up when one of the operators finally got through to him. Know what his first visualization was?"

The visitor shook his head and laughed. "I could try a guess, I suppose," he said, "but my chances would be something less than one in a thousand million."

Moran grinned. "You're so right. There was a whole bunch of kids standing around. Looked like dozens of 'em. And they were all chanting at the top of their voices. You know that old jingle? 'Howie's got a gir-rul?' Chanted it over and over." The grin widened. "Operator said his face stung for ten minutes. That girl must have packed one sweet wallop!"